密斯的隐思
Mies van der Rohe's Secret

王　昀
Wang Yun

中国建筑工业出版社

图书在版编目（CIP）数据

密斯的隐思 / 王昀. - 北京：中国建筑工业出版社，2016.2
ISBN 978-7-112-19066-9

Ⅰ. ①密… Ⅱ. ①王… Ⅲ. ①建筑设计－作品集－德国－现代②德罗（1886～1969）－建筑艺术－建筑理论－理论研究 Ⅳ. ①TU206②TU-80

中国版本图书馆CIP数据核字（2016）第022352号

本教材受北京建筑大学研究生专业实践创新平台项目资助出版

责任编辑：李　鸽　毋婷娴
责任校对：李欣慰　刘　钰
英文翻译：陈伟航
版式设计：张捍平

密斯的隐思

王　昀

*
中国建筑工业出版社出版、发行（北京西郊百万庄）
各地新华书店、建筑书店经销
北京顺诚彩色印刷有限公司印刷
*
开本：889×1194毫米 1/20 印张：$4\frac{2}{5}$ 插页：3 字数：50千字
2016年2月第一版　2016年2月第一次印刷
定价：68.00元
ISBN 978-7-112-19066-9
　　　　（28401）
版权所有　翻印必究
如有印装质量问题，可寄本社退换
（邮政编码 100037）

Mies van der Rohe's Secret

Wang Yun

密斯的隐思

王 昀

Preface 自序

  在建筑设计的过程中,"点子"与"手段"如何结合和使用是一个重要的设计方法论问题。本书从读解密斯在设计巴塞罗那德国馆过程中所绘制的草图入手,直观了密斯本人在设计巴塞罗那德国馆时所运用的"手段"及运用其表达"点子"的过程。为理解密斯在建筑设计过程中"点子"与"手段"的结合与运用方式提供一个思考路径。

  本书分为两个部分,第一部分是以《从巴塞罗那德国馆的建筑平面中读解密斯的设计概念》论文为题,对密斯与赖特的罗比住宅之间的关系进行比对的同时,试图对密斯所设计的巴塞罗那德国馆过程中的"模仿"与"超越"过程进行诠释。而第二部分则是以图片形式来呈现笔者在 1995 年探访巴塞罗那德国馆所看到的景与像,因为这些从探访中所获得的景与像,实际上是我个人对密斯的巴塞罗那德国馆设计过程产生研究与思考兴趣的开始。

During architectural design, it is a significant issue of design methodology regarding how to integrate and use "ideas" and "methods". This book, starting from interpretation of draft drawn by Mies van der Rohe during his designing process of the German Pavilion in Barcelona, demonstrates the "methods" and process of using such methods to express "ideas" when he designed the German Pavilion. It provides a way of thinking to understand integration and usage of "ideas" and "methods" during his architectural design.

The book contains two parts. The first part, with the title of Interpretation of Design Concept of Mies van der Rohe from the Plan of the German Pavilion in Barcelona, compares the relation between Mies van der Rohe and house for Frederick C. Robie, then attempts to illustrate the process of "imitation" and "transcendence" during the designing process of the German Pavilion in Barcelona by Mies van der Rohe. The second part presents scenes and images in the form of pictures, which was observed by the author in 1995 during the visit of the German Pavilion in Barcelona, as these scenes and images from the visit actually ignited the interest to reflect and study the designing process of the German Pavilion in Barcelona by Mies van der Rohe.

王 昀
Wang Yun
2015 年 05 月

	Table of Contents 目录
	Preface 自序
1	Thoughts 思考
3	Interpretation of Design Concept of Mies van der Rohe 从巴塞罗那德国馆的建筑 from the Plan of the German Pavilion in Barcelona 平面中读解密斯的设计概念
25	Site 现场
71	About the Author 作者介绍

1

思考
Thoughts

图 1 Figure 1.

从巴塞罗那德国馆的建筑平面中读解密斯的设计概念
Interpretation of Design Concept of Mies van der Rohe from the Plan of the German Pavilion in Barcelona

建于西班牙巴塞罗那的世界博览会德国馆（图1）是世界建筑大师密斯·凡·德·罗具有代表性的建筑作品，也是他的成名之作。此建筑的出现不仅确定了所谓流动空间的概念，同时也确定了密斯本人在近代建筑史中的地位。关于这一点，建筑史中早有定论，而令我感兴趣的是从这一建筑平面中可以读解出密斯这位大师的构思过程和设计概念。

在德国馆的建筑平面中，有一处一直令人不解的空间构图和平面处理，即在建筑的平面中不知为什么要设置一个由双层玻璃墙围合成的夹层（图2）。因为从整个建筑的平面构图上看，围合室内外空间的墙面是以大理石和玻璃为材料，而且在平面的表示上又是以单线进行。这种以双线表示夹层的做法出于怎样的想法？为了达到怎样的目的？单

Mies van der Rohe earned an international fame from his representative work, the German Pavilion of World Expo held in Barcelona, Spain (figure 1). The pavilion not only defined the concept of fluid space, but also determined the status of Mies van der Rohe in modern architecture history. It was an indisputable fact. Now, what I am interested is to interpret his design process and concept from the plan of the German Pavilion.

In the plan, there was a puzzling spatial structure configuration and graphic design, to be specific, an interlayer enclosed by double glass (figure 2) was employed in the plan. As we can see from the master plan, walls enclosing indoor and outdoor space were constructed in marble and glass while marked on the plan by single lines. What was represented in the idea of expressing the interlayer by double lines? What was the aim? Was this for purely actual function, or for plan composition, or for other reasons? Confused by these questions, I attempted

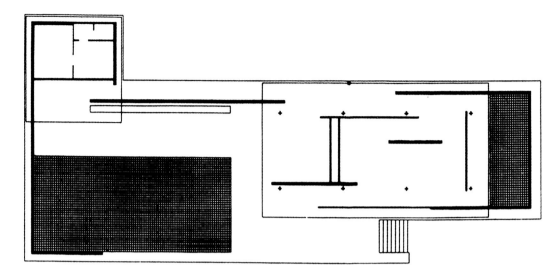

图2 Figure 2.

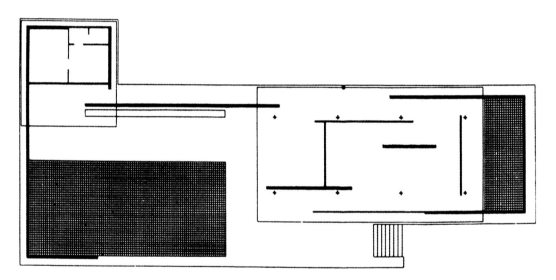

图3 Figure 3.

纯是为了使用功能上的需要，还是出于平面构图上的考虑，或者还有什么别的原因？面对这种种疑问，我也曾试着将这平面中的夹层做过一些改动，例如将夹层去掉，换上单线表示的墙面（图3），但结果的确是让人感到平面的构图中似乎缺少了些什么，显得不那么完美。于是我便得出一个结论，即密斯在设计时为了满足平面构图上的视觉需要才在此设计了这样一个带夹层空间的平面。

几年的时间过去了，两年前一次偶然的机会我来到巴塞罗那，探访了这座著名的德国馆。它是由西班牙政府于1986年在原址上重建的。由于曾对此建筑怀有过诸多的疑问，所以在观览时自然也就对平面上这个夹层部分进行了特别细心的审视。这个夹层在建筑中的处理实际上是以毛玻璃围合而

to alter the interlayer of the structure design, for instance, replacing the interlayer with walls expressed by single lines (figure 3). Somehow as a result, there seemed to be something missing from the plan, leading to imperfection. I therefore reached a conclusion that the plan of space with interlayer was designed to meet visual needs of plan configuration.

Several years elapsed afterwards. Two years ago, I had a chance to visit Barcelona and pay my respect to the well-known pavilion, which was rebuilt by the Spanish Government in 1986 on the original site. With so many questions about this construction, I consequently took a careful examination on the interlayer part. The interlayer, constructed in ground glass, seemed to have no practical function. However, as a skylight was set up right above the interlayer, with special diffused reflection effect by ground glass, the double-glass structure appeared like a bright light box. Since the surrounding indoor area was quite dim, the interlayer itself was highlighted

图4 Figure 4.

成的，虽然看不出有什么使用功能上的需要，但夹层内的上部开有天窗，凭借毛玻璃的漫反射效果，整个夹层形如光箱。加之周围室内空间处理得较暗，实际上夹层本身在整座建筑中显得格外突出（图4）。由于建筑中使用的材料多以不透明的大理石和透明的玻璃为主，所以在这二种对立的建材中出现一种完全不同材质的夹层，不仅与前二者在视觉上形成对比，同时也更加突出了此夹层作为视觉中心在建筑中所起的作用。

通过这样的实地考察和印证，更使我确信了当初的想法，即密斯在此建筑的设计中是刻意地强调这个夹层在整个建筑空间的中心地位及其于平面中的构图作用。为了达到和突出这个目的，他采用了两种手法，一是在平面构图上加以强调

in the entire construction as a unique portion (figure 4). Additionally, as the materials employed in the building was opaque marble and transparent glass, the appearance of the interlayer with a totally different material between two opposed materials not only established a visional contrast with the other two but also emphasized the existence of the interlayer as a visual focus in the entire construction.

After a careful on-the-spot examination and verification, I was convinced of my initial idea that Mies van der Rohe was highlighting the interlayer in terms of its central status in spatial configuration and its configuration function in the plan. To achieve and emphasize this goal, Mies van der Rohe introduced two methods: on one hand, he highlighted the concept in the plan configurations (by using double lines); on the other hand, he emphasized the concept by adopting ground glass.

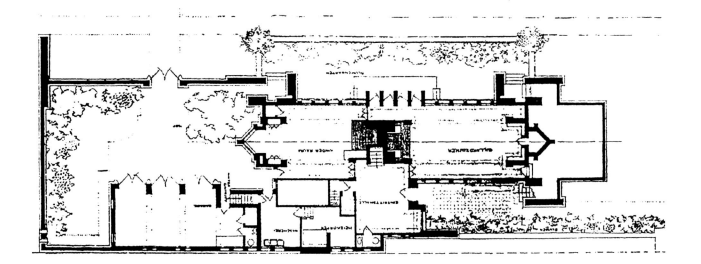

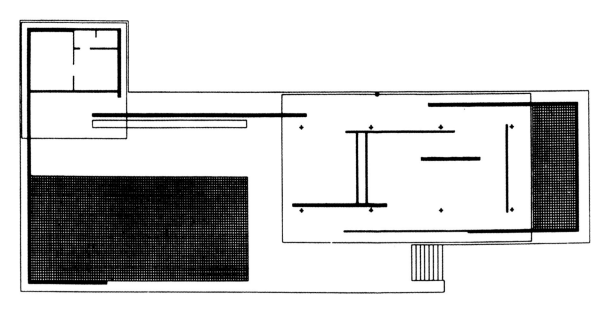

图5 Figure 5.

（采用双线）；二是在材料的运用上加以突出（运用毛玻璃）。

　　至此问题似乎已经得出了答案，但就在几天前偶然翻阅建筑家赖特的建筑作品集时，密斯所设计的这个德国馆的平面又再一次引起了我的兴趣。因为赖特所设计的芝加哥罗比住宅（设计于1909年）的建筑平面从视觉上诉诸于我的是，该平面所呈现出的空间构成与比例和巴塞罗那德国馆平面中所表现出的空间构成与比例具有着某种相互的一致性（图5）。赖特于1909年设计的芝加哥罗比住宅，是他草原式风格建筑中的杰作。该建筑以壁炉为中心，强调水平方向横线条的构成关系，房间各要素的处理强调内部空间和外部空间的互相穿插和流动。从这一建筑中能明显地看出赖特在当时已开始

Thus, it seemed a period could be drawn for the issue. Yet, coincidentally, when I ran into a collection of Frank Lloyd Wright's architecture work several days ago, my interests for the German Pavilion was again provoked. The plan of house for Frederick C. Robie in Chicago (designed by Frank Lloyd Wright in 1909) had certain similarities with the Barcelona German Pavilion in both spatial configuration and scales (figure 5). The Robie House was a masterpiece and classic symbol from Wright's free flowing open space style. Centered on wall furnace, the house design was focused on interrelationships of configuration in the horizontal lines, while handling of different elements of each individual room was focused on inter-combination, inter-conversion and fluidity of interior and exterior space. From this perspective, Frank Lloyd Wright had already employed and attempted the alleged fluid space concept. The house for Frederick C. Robie in Chicago was built in 1909, while the German Pavilion in Barcelona was completed in 1929. With a gap of almost two decades, could

对所谓流动空间的概念进行了运用和尝试。赖特所设计的罗比住宅建于1909年，密斯的世界博览会德国馆建于1929年，二者的年代相差近20年，它们之间是否存在着一个因果关系呢？

  为了解决这个疑问，我试将这两个作品的平面摆放到一起，并重点进行了以下5个方面的比较（图6）。

  1.将两平面缩扩为同样大小并相互比较其相似性。

  2.比较其关键部分空间是否存在着一致性。

  3.墙的位置关系是否存在有重合性。

  4.进出口关系的特征是否一致。

  令人惊奇的是上述比较的结果存在着出人意料的一致性。并从中不难看出二者之间存在有明显

there be a causal relationship between these two buildings?

To solve the problem, I placed the plans of the two buildings and made a comparison primarily in the following 5 aspects (figure 6).

1.To coordinate the plans of the two buildings into the same size to compare similarity.
2.To compare key spatial parts to see if there is any consistency.
3.To compare wall positions to see if there is any possible superposition.
4.To compare interrelationship features of entrance and exit to see if there is any consistency.

Surprisingly, the comparison demonstrated a striking consistency between the two buildings. It went without saying that there was a strong cause-and-effect relationship between the two buildings. Generally, photographers are regarded as the eyes of architects. I

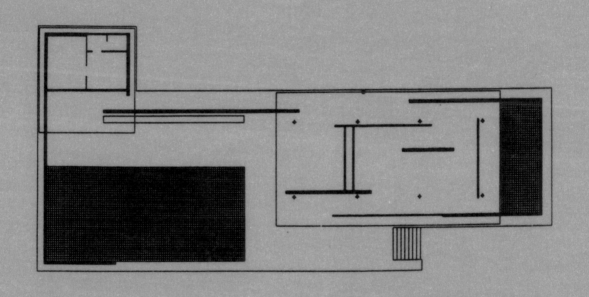

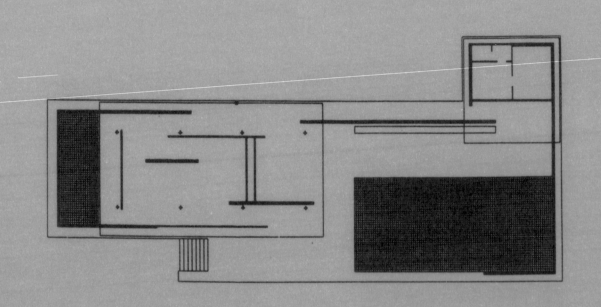

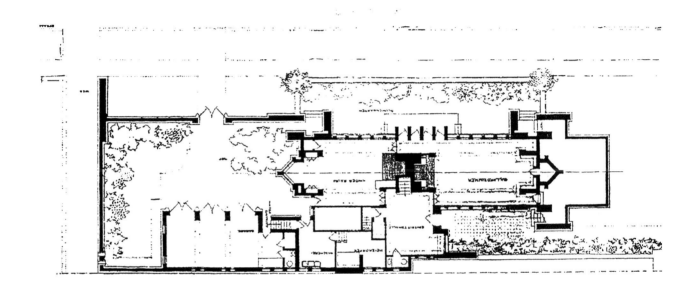

图6 Figure 6

图7 Figure 7.

图8 Figure 8.

因果关系的事实。考虑到建筑摄影家一般又经常地被称为建筑家的眼睛，我有意又将赖特的罗比住宅的建筑图片与密斯的巴塞罗那德国馆的建筑图片加以第5项比较。图7是赖特的罗比住宅的建筑图片，我以为它是赖特观察自己设计的建筑时所采取的视点与角度。图8是巴塞罗那德国馆的建筑图片，同样它也是密斯观察自己设计的建筑时所采取的视点与角度。单纯地从图像现象的角度上看，两者向我们诉说的是纯属相同的结果与理念。特别值得注意的是两张图片在构图比例和平衡关系处理上的相似性，如图7中的树木和图8中的旗杆的位置显得是那么地一致，而且两者在建筑屋顶和墙面之间强调水平关系上所采取的一致性态度又是那么突出。这一切都说明了这两者确实在建筑上存在着相互的关

therefore, purposefully made an additional comparison between the photos of the house and the pavilion in terms of the 5th aspects. Figure 7 is a photo of the Robie House, which I believe to be the viewpoint and angle when Frank Lloyd Wright observed the Robie House designed by himself. Figure 8 is a photo of the pavilion, which was also the original viewpoint and angle adopted by Mies van der Rohe. Simply judging from these graphics, we can find that these two express identical result and conception. It is especially worth noticing that there is a similarity as to the configuration scales and balancing connections of the two photos. As an example, the tree position in figure 7 and flagstaff position in figure 8 appear to be evidently consistent, and it is obvious that two buildings adopt a same designing attitude emphasizing horizontal relationships between roof and wall. All of these factors illustrate that these two buildings do exist a causal relationship. So far from the phenomenon, we have found coherency from these buildings, yet, is this association an accidental coincidence or a consequential affiliation?

联。分析至此，我们已经从现象中找出了两者之间的一致性。但两者之间是纯属偶然的存在，还是有一种必然的关联？

事实上，1910年代的赖特在欧州曾享有过很高声望，并且有过很大的影响。密斯在1940年前后曾对赖特有过这样的评价，他说："这位巨匠的作品中有意想不到力量……他的作品越仔细地研讨，越感到他那大胆的构想和无比的才能，越觉得值得对他那独自的思考力和行动给予高度的赞扬，从他作品中所放出的生气勃勃的冲击力鼓舞了所有的世代。他的影响甚至即使其作品不在眼前也会紧紧地真实地感到。"[1]这段论述的文字中有这样特别值得注意的几句话，首先密斯所说的"他（赖特）的作品越仔细地研讨"这句话本身就意味着密斯曾对赖

In fact, Frank Lloyd Wright enjoyed a high reputation and influence in Europe during 1910s. Mies van der Rohe had the following comments on Frank Lloyd Wright around 1940s. He said: "This great grandmaster has an unimaginable power……. The more I delve into his works, the more I feel that he has such a bold imagination full of unparallel intelligence, while his independent thinking and unique work deserve supreme acclaims. The vigorous impact from his works inspires all eras and his influence, even in absence of his works, can be felt closely and vividly."[1] Special attention should be paid on the following sentences. In the first instance, Mies van der Rohe claimed: "The more I delve into his works", which implied a fact that Mies van der Rohe had studied Frank Lloyd Wright's works in a meticulous manner. Subsequently, Mies van der Rohe mentioned that the vigorous impact was inspiring and his influence, "even in absence of his works, can be felt closely and vividly". All these words manifested that Frank Lloyd Wright's design concept had such a tremendous

特的作品作过仔细研讨的事实,并且他接着又说:对赖特的作品越仔细地研讨就越从中感到生气勃勃,即使赖特的"作品不在眼前也会紧紧地真实地感到"等等。所有这些都表现出密斯受赖特的影响之大,以至于即使赖特的作品不在其眼前也可以真实地感觉到,可见赖特的作品已深深地印在了密斯的脑海里。

同样的另一个耐人寻味的印证是,密斯曾确确实实崇拜过赖特。在由山本学治和稻葉武司共著的《巨匠密斯的遗产》一书中,曾对1910~1920年这一时期密斯的设计思想作过这样的分析和解释,他们认为当时对密斯设计思想影响最大的是1910年在柏林举办的赖特建筑作品展。关于这次展览,密斯曾这样谈到他的感受,他说"我们年轻的建筑家

influence on Mies van der Rohe that even without seeing an actual Frand Lloyd Wright's works, Mies van der Rohe could feel Frank Lloyd Wright's design vividly. Obviously, Frank Lloyd Wright's design was deeply engraved in Mies's mind.

Another interesting evidence was that Mies van der Rohe did actually adore Frank Lloyd Wright. In the book of "The Heritage of Great Architects"published by Japanese scholars (Takeshi Yinabaand Gakuji Yamamoto), analysis and interpretation of Mies van der Rohe's design concepts during the period of 1910~1920 was given. They believed that the greatest influence exerted on Mies van der Rohe was the exhibition of Frank Lloyd Wright's works in 1910 held in Berlin. Regarding to this exhibition, Mies van der Rohe had following remarks: "Our young architects are trapped in drastic mental confusion. However, the works of this master architect not only present striking varieties and clarity in terms of exterior appearance, but also unfold an architectural world

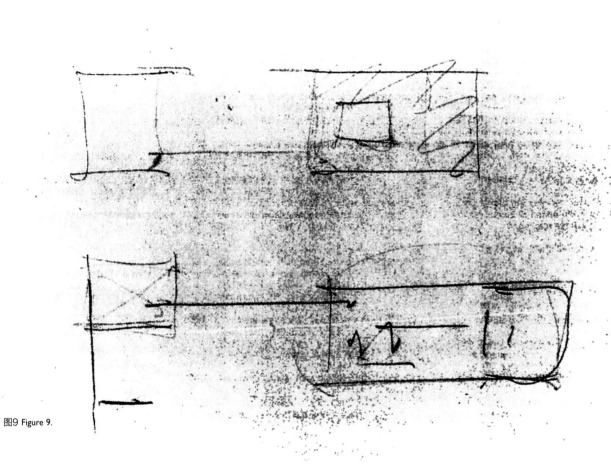

图9 Figure 9.

们，正陷入在急剧的精神混乱中。然而这位巨匠的作品，不仅在造型上表现出了令人瞠目的丰富性与明澈性，同时还展现给我们一个充满着意想不到力量的建筑世界。"短短的几句话，赖特对密斯的决定性的影响已表述得淋漓尽致。因为这句话表明了赖特给当时正陷入在"急剧的精神混乱中"的密斯"展现了一个充满着意想不到力量的建筑世界"。

正值本文即将下结论时，我又发现另一个更有说服力的证据，那是一张被认为是密斯在设计巴塞罗那德国馆时所绘的第一张草图（图9），草图上表现了德国馆在初始构思时就与罗比住宅的平面有着直接的关系。如将两者相互比较，可以清楚地看出密斯最初留在草图上那借鉴赖特设计罗比住宅时的痕迹（图10~图12）。

full of unimaginable power." The determining influence of Frank Lloyd Wright on Mies van der Rohe was exposed fully within these few sentences. In fact, these words conveyed that Mies van der Rohe was in a "drastic mental confusion" while Frank Lloyd Wright's works gave him an architecture world full of unimaginable power.

Another convincing evidence appeared when I was about to make the conclusion: the first draft (figure 9) used by Mies van der Rohe, as it was identified, when he designed the German Pavilion in Barcelona. From the draft, a direct relationship between the house and the pavilion during initial design stage can be seen. If we compare these two, we can see a clear trace that Mies van der Rohe made a reference to Frederick C. Robie from the initial draft (figure 10~figure 12). So far, we now may draw a conclusion that design of the pavilion was developed from the house for Frederick C. Robie

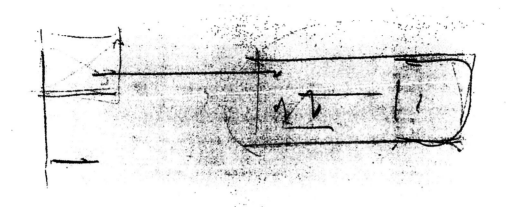

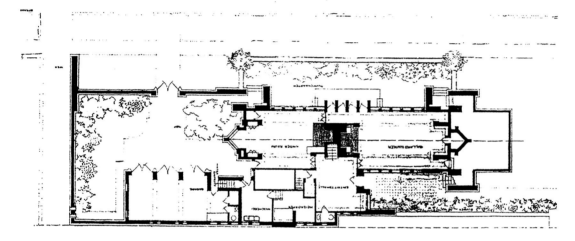

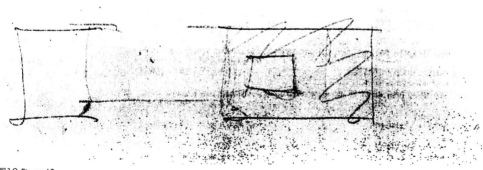

图10 Figure 10.

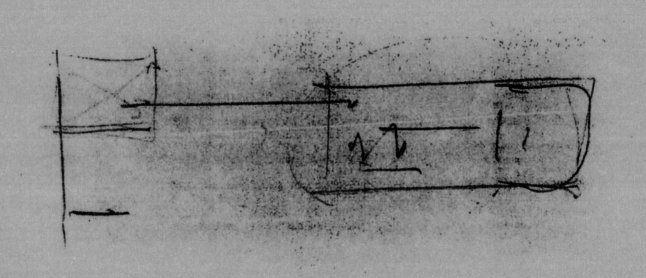

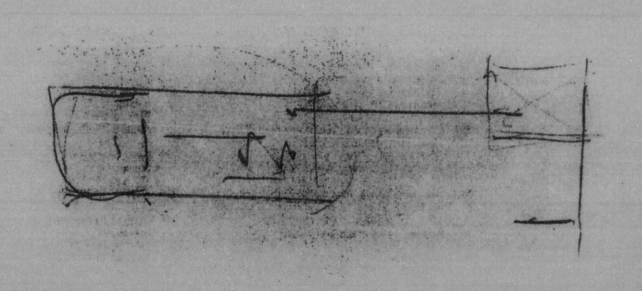

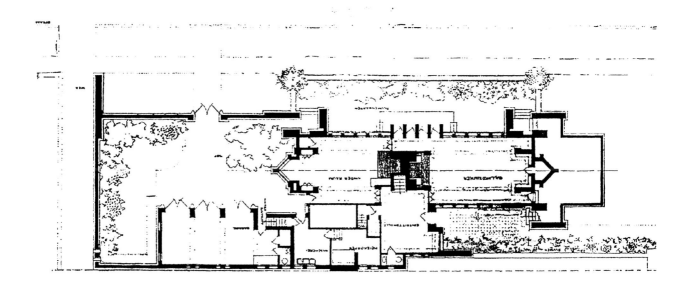

图11 Figure 11.

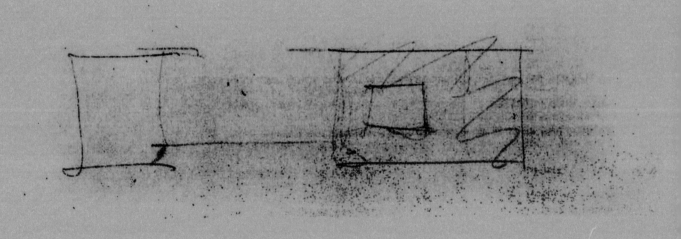

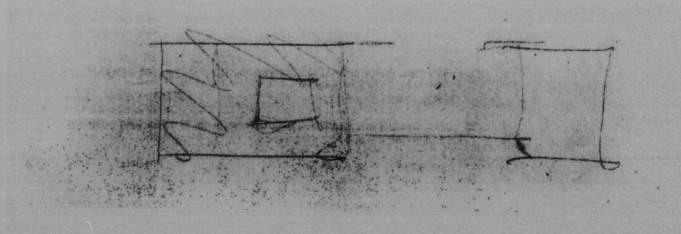

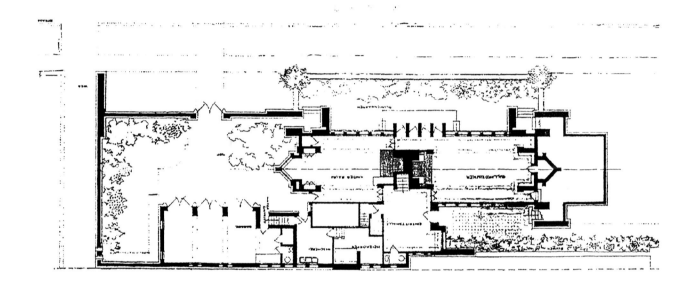

图12 Figure 12.

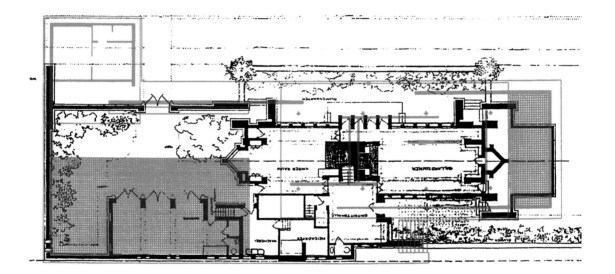

图13 Figure 13.

到此为止终于可以下这样的结论：即密斯在巴塞罗那德国馆中所表现的设计概念和意图是以赖特的罗比住宅为蓝本发展而来的。但我以为密斯所追求的在空间与造型上的纯粹性，与赖特当时的装饰性和尚未完全脱胎换骨的古典性有着根本的区别。密斯的伟大正是在于他勇于运用抽象的概念，勇于去掉复杂，勇于从本质上表达建筑，正是因为这个纯粹性从而使得密斯的建筑更明晰地表达了赖特曾要表达的思想和理念。

我以为，密斯在设计巴塞罗那德国馆时，是以赖特设计的罗比住宅平面为草底，并在上面进行构草。但技巧出众的是，他勇于减掉更多复杂的东西，使复杂变为单纯。他去掉了赖特建筑中的装饰，因为他认为装饰就是罪恶；他去掉了赖特建筑中那些多余的封闭墙，那是为了表现少就是多。最后我也在这样的分析过程中得到了我对巴塞罗那德国馆夹层处理的那一不解之迷的答案，因为那是赖特设计的罗比住宅中的壁炉的位置（图13）。

as a blueprint. However, in my opinion, Mies van der Rohe's pursuit was seeking simplicity and quintessence in terms of space and exterior appearance, which was fundamentally different from Frank Lloyd Wright's ornamental and immature classic style. The greatness of Mies van der Rohe is that he was brave to express and convey the nature of architectures with abstract concepts rid of complexity. It was the simplicity and quintessence that enabled Mies van der Rohe to express, in a clear manner, idea and concept which Frank Lloyd Wright attempted to convey in his architectures.

I believe,that Mies van der Rohe designed the pavilion on the basis of the sketch of house of Frederick C. Robie and then developed from it. Having an outstanding skill, he bravely abridged the complexity and achieved simplicity. He wiped off the decorations of Wright's, because he considered ornaments as crimes; he got rid of the extra obturate walls from Wright's house, because less is more. Eventually, with the analysis above, I find the answer for the long-time unsolved puzzle about the interlayer in the German Pavilion in Barcelona, which happens to occupy the position where the furnace was in Robie House (figure 13).

2

现场
Site

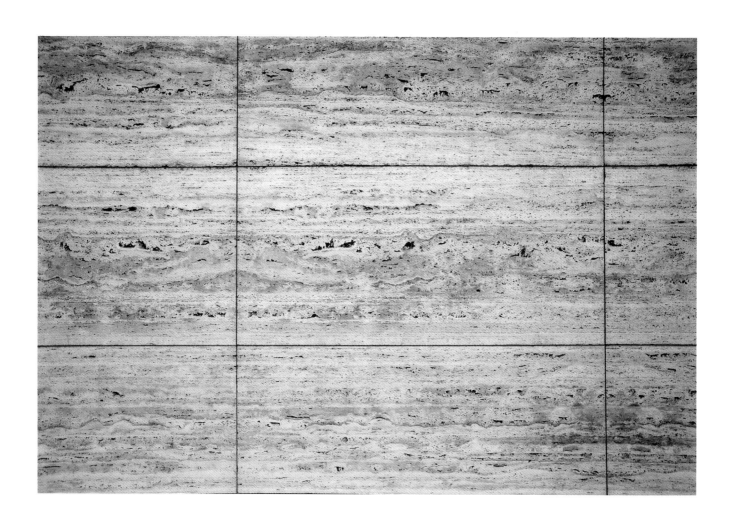

61

图片说明：

24　从1929年世博会西班牙国家宫前，望向北侧的大台阶及喷泉，密斯为此次世博会设计的德国馆位于本图片中喷泉的左侧。

26　从喷泉的西侧人行道旁看到德国馆的东立面

27　站在喷泉广场远望德国馆

28　在水池前正对德国馆的东立面

29　沿人行步道接近德国馆的视角

30　德国馆朝北向的入口

31　从南向看到的德国馆与周边环境

32　站在街道上看德国馆的水池

33　入口的楼梯与玻璃窗后面的红色窗帘

34　从室内看到北向室外的雕塑与水池

35　从室内向北望向室外的庭院

36　由磨砂玻璃所围合的德国馆的室内"光箱"，这个"光箱"成为引发本论的发想源头。

37　透过玻璃望向庭院

38　庭院西北向的水池与雕塑

39　站在庭院，从东向西看到的庭院西立面

40　室外庭院的东北向视角

41　室外庭院的东立面

42　作为视觉趣味中心的室外雕塑

43　在西侧入口穿过室内望向北侧的室外庭院

44　西侧入口的南向场景

45　屋顶、墙、柱子与地面的组合关系

24　Standing in front of the Spain National Palace built for World Expo 1929 and looking at big steps and fountain on the north side, we can see the German Pavilion designed by Mies van der Rohe for this World Expo is located on the left side of the fountain in this picture

26　East facade of the German Pavilion seen from sidewalk west to the fountain

27　Distant view of the German Pavilion seen from Fountain Square

28　East facade of the German Pavilion right opposite to the pool

29　Perspective as one approaches the German Pavilion along sidewalk

30　Northward entrance of the German Pavilion

31　The German Pavilion and its surroundings seen from south

32　Pool of the German Pavilion seen from street

33　Steps at entrance and red curtain behind glass window

34　Outdoor sculpture and pool seen indoors northward

35　Outdoor courtyard seen indoors northward

36　Indoor "light box" enclosed by ground glass in the German Pavilion, which triggers this subject

37　Courtyard seen through glass

38　Pool and sculpture northwestward in courtyard

39　West facade of courtyard when one stands inside it, viewing from east to west

40　Northeastward perspective of outdoor courtyard

41　East facade of outdoor courtyard

42　Outdoor sculpture as visual interest center

43　Passing indoor space from west entrance and looking at outdoor courtyard on north side

44　Southward scene at west entrance

45　Combination relation of roof, wall, pillar and ground

#	中文	#	English
46	墙面、地面与屋顶的搭接关系	46	Connection relation of wall, ground and roof
47	由西侧所看到的德国馆东南角的水池	47	Pool on southeastern corner of the German Pavilion seen from west side
48	在东侧入口望向南侧的水池	48	Pool on south side seen from east entrance
49	水池西侧的座椅	49	Bench on west side of pool
50	入口的西南向场景	50	Southwestward scene at entrance
51	对水池进行空间限定的墙	51	Wall to delineate pool space
52	从西南角向东北方向的场景	52	Northeastward scene from southwest corner
53	从南边的围墙前,向北所看到的场景	53	Northward scene seen in front of south wall
54	在水池旁向北所看到的场景	54	Northward scene seen beside pool
55	从室外看到的磨砂玻璃所围合的德国馆的"光箱"	55	The "light box" enclosed by ground glass in the German Pavilion seen outdoors
56	从水池东向西侧所看到的场景	56	Scene seen at pool from east to west
57	从水池东向北侧所看到的场景	57	Scene seen at pool from east to north
58	石材墙面的肌理	58	Texture of stone wall surface
59	室内石材墙面与摆放在前面的密斯设计的座椅	59	Indoor stone wall surface and chair in the front designed by Mies van der Rohe
60	室外墙面	60	Outdoor wall surface
61	柱子、墙体与屋顶的结合关系	61	Combination relation of pillar, wall and roof
62	水池的细部	62	Pool details
63	室外地面的细部	63	Outdoor ground details
64	隔墙、柱子与地面的细节	64	Partition wall, pillar and ground details
65	从德国馆西侧的台地上能够看到屋顶上突起来的由磨砂玻璃所围合的"光箱"顶的天窗	65	From the terrace on the west side of the German Pavilion, one can see on the roof jutting skylight of "light box" enclosed by ground glass
66 67	德国馆的西侧场景	66 67	Scene on the west side of the German Pavilion

作者介绍
About the Author

王 昀 简介

王 昀 博士
1985年毕业于北京建筑工程学院建筑系，获学士学位
1995年毕业于日本东京大学，获得工学硕士学位
1999年于日本东京大学获得工学博士学位
2001年执教于北京大学
2002年成立方体空间工作室
2013年于北京建筑大学创立建筑设计艺术研究中心

建筑设计竞赛获奖经历：
1993年日本《新建筑》第20回日新工业建筑设计
　　　竞赛获二等奖
1994年日本《新建筑》第4回S×L建筑设计竞赛
　　　获一等奖

主要建筑作品：
善美办公楼门厅增建、60平方米极小城市、石景山财政局培训中心、庐师山庄、百子湾中学校、百子湾幼儿园、杭州西溪湿地艺术村H地块会所等

参加展览：
2004年6月参加"'状态'中国青年建筑师8人展"
2004年首届中国国际建筑艺术双年展参展
2006年第二届中国国际建筑艺术双年展参展
2009年参加在比利时布鲁塞尔举办的"'心造'——中国当代建筑前沿展"
2010年参加威尼斯建筑艺术双年展、德国karlsruhe Chinese Regional Architectural Creation 建筑展
2011年参加捷克prague中国当代建筑展、意大利罗马"向东方——中国建筑景观"展、中国深圳·香港城市建筑双城双年展等
2012年第13届威尼斯国际建筑艺术双年展中国馆参展

Dr. Wang Yun
Graduated with a Bachelor's degree from the Department of Architecture at the Beijing Institute of Architectural Engineering in 1985.
Received his Master's degree in Engineering Science from Tokyo University in 1995.
Received a Ph.D. from Tokyo University in 1999.
Taught at Peking University since 2001.
Founded the Aterier Fronti (www.fronti.cn) in 2002.
Established Graduate School of Architecture Design and Art of Beijing University of Civil Engineering and Architecture in 2013.
Prizes:
Received the second place prize in the "New Architecture"category at Japan's 20th annual International Architectural Design Competition in 1993.
Awarded the first prize in the "New Architecture"category at Japan's 4th S×L International Architectural Design Competition in 1994.

Prominent works:
ShanMei Office Building Foyer, a Small City of 60 Square Meters, the Shijingshan Bureau of Finance Training Center, Lushi Mountain Villa, Baiziwan Middle School, Baiziwan Kindergarten, and block H of the Hangzhou Xixi Wetland Art Village.
Exhibitions:
The 2004 Chinese National Young Architects 8 Man Exhibition, the First China International Architecture Biennale, the Second China International Architecture Biennale in 2006, the "Heart-Made: Cutting-Edge of Chinese Contemporary Architecture" exhibit in Brussels in 2009, the 2010 Architectural Venice Biennale, the Karlsruhe Chinese Regional Architectural Creation exhibition in Germany, the Chinese Contemporary Architecture Exhibition in Prague in 2011, the "Towards the East: Chinese Landscape Architecture" exhibition in Rome, and the Hong Kong-Shenzhen Twin Cities Urban Planning Biennale. The thirteen Venice Architecture-Art Biennale in 2012.

注释：
[1]《巨匠ミースの遺産》山本学治，稲葉武司 共著，彰国社刊

Footnote:
[1] "The Bequest of Great Architects" by Takeshi Yinaba and Gakuji Yamamoto, Shokokusha Publishing Co., Ltd.

说明：
[1]《从巴塞罗那德国馆的建筑平面中解读密斯的设计概念》一文原刊载于《华中建筑》2002年01期第20卷
[2] 本书"现场"一章中的图片均由作者拍摄于1995年10月

Note:
[1] "Interpretation of Design Concept of Mies van der Rohe from the Plan of German Pavilion in Barcelona" was initially published on Central China Architecture Issue 1, Vol. 20 in 2002
[2] Photos in the chapter of "Site" of this book were all taken by the author in October 1995

参考文献 | Reference：
[1] "The Mies van der Rohe Archive" Vol. 2 Edited by Arthur Drexler

关于版权：
由于相关图纸没有找到版权的所有者，如果版权所有者看到此书，请及时与作者联系，以便支付使用图纸所产生的费用。

Copyright:
As copyright owners of related drawings are not found, they are kindly asked to contact the author in time if they read this book so that we can offer them payment for using these drawings.